Immersive Technologies In Healthcare: Virtual Reality And Augmented Reality

Mbuso Mabuza

Published by Mbuso Mabuza, 2024.

IMMERSIVE TECHNOLOGIES IN HEALTHCARE: VIRTUAL REALITY AND AUGMENTED REALITY

First edition. February 1, 2024.

ISBN: 979-8215811450

Written by Mbuso Mabuza.

Also by Mbuso Mabuza

A Healthy Mind And Best You: Achieving Great Results in Every Aspect of Your Life

Purposeful And Better You

Sustainable Development Calls for Effective Strategic Leadership for Efficient Health Systems

Health Promotion In Low Socioeconomic Settings

Medicine and Sociology of Health

Qualitative Methods In Public Health Research

Global Health Disaster Management

Global Health Policy And Programme Challenges

The Journey of Life Has a Gift of Purpose

Epidemiological Research

Ethics, Qualitative And Quantitative Methods In Public Health Research

When Love Lasts

Blockchain Technology In Healthcare And Medicine

Immersive Technologies In Healthcare: Virtual Reality And Augmented Reality

Health Data Analytics And Informatics

How To Improve The Way You Think

Health Systems Engineering: Building A Better Healthcare Delivery System

Artificial Intelligence In Drug Discovery And Development

How To Make The Most Of Life

Precision Medicine

Connected Health: Technology-Enabled Care

Table of Contents

Preface

Virtual reality (VR) and augmented reality (AR) are being increasingly used in healthcare, and especially in patient education. Although these two technologies are very similar, virtual reality engages users by transposing their virtual avatars into an artificial environment for users to interact with, while augmented reality engages users in the real environment by overlaying virtual elements in the real world.

VR is already transforming medical education. VR offers distinct benefits for learners, faculty and the health system. It is helping to free learning from the classroom, allowing learners to apply their knowledge to practise and learn from mistakes. It focuses on improving competencies and places the emphasis on autonomous, blended learning, which is expected from the learners of today.

Healthcare and medical staff can use virtual environments to train in everything from surgical procedures to diagnosing a patient. VR training curriculums were created in order to allow both training and practising surgeons to a safer operating room experience for their patients. VR can be incorporated into a surgeon's preoperative planning, utilised in unison with artificial intelligence (AI) algorithms, to help virtually map a procedure.

VR is being incorporated into occupational therapy to help stroke patients recover. VR-based rehabilitation is a proven tool that creates specific scenarios for patients, allowing them to have targeted treatments for their particular recovery level and deficits.

Off-shelf virtual reality technology using simple headsets has promise for the prevention and treatment of health conditions, particularly in psychological care. For mental health issues such as dementia, depression and stress management, this technology provides an effective and better solution which enhances patient and creates a positive effect to save the life of the patient.

Virtual patients are seen as cost-effective since virtual patients are limiting the effort and expense associated with standardised patient training and can be delivered at low-cost over the internet. There is potential for virtual healthcare to address inequalities in healthcare. It can bridge the gap for marginalised and vulnerable populations living in remote settings by increasing healthcare access for children and families who may otherwise face significant barriers in having their health needs met. The main challenges of developing virtual patents are legal and technical.

With augmented reality (AR), there is no created scenario; instead, an actual event is being altered in real-time, which has significantly enhanced robotic surgery. AR can work in parallel with a telemanipulation system in order to optimise the visual field of the operating surgeon. AR is currently being utilised to overlay key anatomic landmarks during live surgery to optimise patient safety.

This book presents an account of virtual reality and augmented reality and their impact on the healthcare and medical ecosystem in the context of low-, middle- and high-income countries. The book concludes by providing future perspectives and critical reflections on VR and AR in healthcare and medicine. It also highlights the new computational medical extended reality (XR) concept, which unifies the computer science applications of intelligent reality, medical virtual reality, medical augmented reality and spatial computing for medical training, planning and navigation content creation.

The book will be invaluable to healthcare and global health professionals, health managers and programmers, health systems leaders, researchers, lecturers in faculties of health sciences, medical students, students in public health and related fields, healthcare organisations, and policymakers.

Chapter 1
Overview of Virtual and Augmented Reality

"A better reality is what we all look for. Our wishes, dreams, desires, aspirations and anything else that, we think, would make us feel good, they all are representations of an intangible realm that we would all like to have. These days, we talk a lot about virtual "Reality" and augmented "Reality". In a sense, the terms are becoming somewhat interchangeable, although the initial definitions had a more strict description... It all might look like science fiction, but it is not. It is intuitive. We already have the technology and capabilities to create such a system, and to make it functional very soon. It does require one to be bold, to embrace innovation. It is not about the technology, but about the idea behind the use of that technology where the genius resides - virtually" – **Dr Rafael J. Grossman, 2015.**

1.1 Introduction

ADVANCES IN DISPLAY technology and computing have led to new devices capable of overlaying digital information onto the physical world or incorporating aspects of the physical world into virtual scenes. These combinations of digital and physical environments are referred to as extended realities (Andrews et al, 2019). Extended reality (XR) technology includes virtual reality (VR), augmented reality (AR) and mixed reality. XR is now widely used in Human Computer Interaction

(HCI), social science and psychology experimentation (Ratcliffe et al, 2021).

The first device that looks like a modern VR headset was invented in the 1830s during the early days of photography. It was called a *stereoscope* and used a slightly different image for each eye to create a 3D effect. The impact of the stereoscope in the mid-nineteenth century has often been likened to that of the television in the 1950s. The devices were incredibly popular, being present in most middle and upper-class drawing rooms, as well as in exhibitions and fairs. However, this golden age of the stereoscope ended abruptly in the 1860s with the advent of the more portable and convenient *'carte-de-visite'* – a type of small photograph.

Just as early photography innovation had initially embraced the solo-viewing experience, so did early film. Like the Stereoscope, Edison's *Kinetoscope* had a lot in common with contemporary VR. It was only when the one-person-at-a-time model was found to be not commercially viable, that group cinema culture emerged in the early 1900s.

Let us now fast forward to the 1960s – to the early days of computing. By now, there had already been various philosophical and fictional writings about virtual and augmented reality, so the concept was firming up in the public imagination. MIT computer scientist Ivan Sutherland set out to create a head-mounted AR machine which he called *The Sword of Damocles*. It allowed users to step into the data; a concept we would now call 'spatial computing'. Although The Sword of Damocles never was released to the public, Sutherland continued his work in the development of computer graphics. In fact, many in the video gaming industry would today describe him as the father of computer graphics.

A couple of decades later in the 1980s, 'tech' culture was really starting to heat up, with a particular hotspot around the San Francisco Bay area. Not only was the region home to companies like IBM, Microsoft and Atari – but there were venture capitalists keen to invest.

The Silicon Valley money, plus an entrepreneurial punk spirit heralded what is now described as AR and VR's first wave. Several teams set about to make virtual reality headsets people could actually buy. One of the most influential companies was VPL Research. The company created a whole virtual reality system, including bodysuit, headset and gloves. The dream was becoming a reality; experimental uses ranged from VR for NASA training to ground-breaking immersive art experiences. In 1989, Angels premiered, the world's first virtual reality movie, directed by Nicole Stenger, and in 1990 Brenda Laurel wrote the seminal text, *Computers as Theatre*.

A generation of other headset makers and artistic pioneers continued to push boundaries. By the mid-1990s, some manufacturing companies were even using VR and AR in their processes. Video game arcades began to include VR machines, Nintendo released their own home VR headset and there was even a major augmented reality dance production performed in Australia, called Dancing in Cyberspace, directed by Julie Martin.

Inventors had got closer than ever before to creating immersive worlds and their attempts had significantly helped develop a new medium of computer graphics. By the mid-2000s, AR was barely known, and VR was perceived as one of those inventions that 'just never took off'. Where to from there?

Luckily, processor speeds were increasing at an exponential rate along with internet speeds. VR, was being used in the giant life simulation platform *Second Life*. AR was given a major boost in the early 2010s with the popularisation of the smartphone: OR codes became commonplace and popular AR app platforms like *Blippar* launched.

Once again, entrepreneurs were beginning to see the opportunities in AR and VR. In 2013, technologist Jeri Ellsworth founded Technical Illusions, and successfully crowdfunded the AR/VR device *CastAR*. There was clearly demand for her invention: it surpassed its funding goal two days after the project went live, raising over $1 million, despite company problems later down the line that resulted in the funding being returned. At a similar time, student Palmer Luckey also launched a crowdfundiing campaign for his own device – a VR headset called an *Oculus*. It also achieved well above its funding target.

Larger tech companies were well aware of this increase in consumer interest and rapid innovation progress. 2014 was a landmark year: Palmer Luckey sold Oculus to Facebook for $3 billion and Google released their AR glasses, Google Glass to the public. Google Glass never took off (after much public mockery), however other tech giants including Samsung, Microsoft, Sony and HTC also got to work developing AR and VR headsets. By 2016, there was a whole suite of AR and VR headsets available on the market.

As for AR on smartphones, often without even realising it, mainstream smartphone users had been regularly experiencing augmented reality, with the hugely popular releases of Snapchat's lens filters and *Pokemon Go*.

We are now in the thick of this wave of virtual and augmented reality. Millions of people actually use immersive technology every day. The conditions are in place for it to finally go mainstream and be considered a media form that sits comfortably alongside film, photography, radio, websites and books. An industry around this medium is in the process of being built and it really is all to play for (Allen, 2021).

1.1.1 Virtual Reality

VIRTUAL REALITY (VR) is a technology which allows a user to interact with a computer-simulated environment, whether that environment is simulation of the real world or an imaginary world. It is the key to experiencing, feeling and touching the past, present and the future. It is the medium of creating our own world, our own customised reality. With virtual reality, we can experience the most intimidating and gruelling situations by playing safe and with a learning perspective (Mandal, 2013).

According to PwC (2019), virtual reality immerses users in a fully digital environment through a headset or surrounding display. This environment can be computer-generated or filmed in 360-degree video. Daily, this can be used in video games, design planning, and even simulated roller coaster rides.

There are countless uses for virtual reality in healthcare and medicine. For example, have you heard about IrisVision? This is a new technology created by Dr Frank Werblin in association with CitrusBits. It is a virtual reality product that restores eyesight to people who have certain diseases of the eye. Whether the patient has optic nerve damage, macular degeneration, or another low-vision condition, this product could help. Another product that makes use of virtual reality is OSSO VR, a product that benefits surgeons and their medical staff. This virtual reality device lets them practice sensitive surgeries on virtual "patients." What's more, these simulations are meticulously realistic even though they happen to be in a training environment. In other words, OSSO VR lets surgeons experiment with new technologies without harming anyone, or worse, causing a fatality.

In healthcare, VR could include a cadaver to help learn anatomical structures and any preoperative imaging to help plan a procedure. VR training curriculums were created in order to allow both training and practising surgeons to a safer operating room experience for their patients. VR can also be incorporated into a surgeon's preoperative planning, utilised in unison with AI algorithms, to help virtually map a procedure. VR is not limited to the operating room; VR is being incorporated into occupational therapy to help stroke patients recover. VR-based rehabilitation is a proven tool that creates specific scenarios for patients, allowing them to have targeted treatments for their particular recovery level and deficits (Moawad et al, 2020).

IMPACT ON THE WORLD

Bring the world closer together. VR will help connect the world and give people visual and immersive opportunities (PwC, 2019b).

Impact on education

Institutions are using VR to augment programmes in criminal science, healthcare, agriculture and fine arts. Through VR, students can practise and train in their field of study. Students studying history could explore landmarks and historical sites; anatomy students could identify and practise procedures on virtual human bodies; and students studying education could practise teaching with the use of virtual students to become more comfortable in the classroom (PwC, 2019b).

Where this has been successfully used

- Western University of Health Sciences lets students learn about anatomical functions by moving layers of virtual tissue to view more than 300 anatomical visualisations, created using scans of real patients and cadavers.
- Harvard University – Virtual Immersion: French Culture and Language through 3D Video. In Nicole Mills' French language and culture classes, students meet native speakers at parties in their homes and eavesdrop on conversations in Parisian cafes, all without leaving the classroom. Cultural immersion is a tried-and-tested element of language instruction, and this project brings students into the 11th arrondissement of Paris through VR film narratives.
- University of Westminster built a virtual space for criminal law students, in which they hunt for clues to construct a murder case. Rather than simply reading witness statements, they can walk around a building and judge whether someone would have been able to see the crime (PwC, 2019b).

VIRTUAL REALITY AND augmented reality are being increasingly used in healthcare, and especially in patient education. Although these two technologies are very similar, virtual reality engages users by transposing their virtual avatars into an artificial environment for users to interact with, while augmented reality engages users in the real environment by overlaying virtual elements in the real world. Due to the immersive nature, virtual reality presents several disadvantages to its users who may feel cyber sickness through disorientation, headache, nausea and other symptoms associated with motion sickness. Additionally, virtual reality users face technological limitations in manipulating tools and components in the virtual world resulting in distraction in learning due to interference with virtual reality. On the other hand, augmented reality removes these limitations in learning by blending into the users' reality with minimal interference; that is, the users do not completely immerse into virtual space and can see the superimposed objects within the real-world environment. This enables the augmented reality users to be more interactive with the task in hand (Adapa et al, 2020).

Virtual reality's impact on the world is that it brings the world together. Virtual reality will help connect the world and give people visual and immersive opportunities. Augmented reality's impact on the world is that marketing and advertisement fields will explode with augmented reality devices. Augmented reality is the future that will allow consumers to experience a reality that is based on personal needs and desires. Augmented reality will present a completely new way to engage and will expand the abilities of retailers as well (PwC, 2019b).

Virtual reality is considered to have begun in the 1950s but it came to the public's attention in the late 1980s and 1990s. This can be attributed to pioneering computer scientist Jaron Lanier who introduced the world to the term 'virtual reality' back in 1987. Most virtual reality environments are primarily visual experiences, displayed either on a computer screen or through special stereoscopic displays. Virtual reality may also include auditory stimulation through speakers or headphones. Users can interact with the virtual environment through the use of devices such as a keyboard, a mouse, or a wired glove. The majority of historical examples of virtual reality are visual and to a lesser extent, auditory. This is because of all the human senses, vision provides by far the most information, followed by hearing (Cipresso et al, 2018; Mandal, 2013).

The global virtual reality market size is projected to reach US$120.5 billion by 2026, exhibiting a CAGR of 42 percent during the forecast period. Explosion of the COVID-19 pandemic is set to open up several growth avenues for the market, suggests Fortune Business Insights in its report, titled "Virtual Reality Market Size, Share & Industry Analysis, By Offering (Hardware, Software), By Technology (Nonimmersive, Semi-Immersive), By Industry Vertical (Gaming & Entertainment Media, Healthcare, Education, Automotive, Aerospace & Defense, Manufacturing), By Application (Training & Simulation, Educational, Attraction, Research & Development) and Regional Forecast, 2019-2026". The coronavirus pandemic has forced people to work from home and follow strict special distancing norms. Virtual reality is one technology that is playing a critical role in supporting businesses to perform their operations as smoothly as possible. The growth of this market, therefore, is set to be augmented as the coronavirus continues its global rampage (Fortune Business Insights, 2020).

Virtual Healthcare is a broad term encompassing all the modes of diagnosing and treating patients without being there. There are countless uses for virtual reality in healthcare and medicine. For example, have you heard about IrisVision? This is a new technology created by Dr Frank Werblin in association with CitrusBits. It is a virtual reality product that restores eyesight to people who have certain diseases of the eye. Whether the patient has optic nerve damage, macular degeneration, or another low-vision condition, this product could help. Another product that makes use of virtual reality is OSSO VR, a product that benefits surgeons and their medical staff. This virtual reality device lets them practise sensitive surgeries on virtual "patients." What's more, these simulations are meticulously realistic even though they happen to be in a training environment. In other words, OSSO VR lets surgeons experiment with new technologies without harming anyone, or worse, causing a fatality.

Virtual reality is fast gaining traction in the healthcare sector, especially in training surgeons. A study published in Harvard Business Review stated that surgeons trained through virtual reality experienced a 230 percent boost in their performance and carried out surgeries with greater precision compared to their conventionally trained counterparts. Even more promising, a study published by the National Institutes of Health showed that virtual reality therapy led to better mobility outcomes in children afflicted by cerebral palsy. Thus widening applications of virtual reality are bolstering the virtual reality market growth (Fortune Business Insights, 2020).

In health education, learning involves multifaceted physiological systems, developed adaptive expertise, and the acquisition of collaborative skills. Learning in the medical domain is often situated in a real-life context yet training in this real-life context is not always possible. To truly achieve simulation-training that replicates medical environments, there needs to be not only analogous mental challenges, but also analogous physiological responses. To replicate and train for human factors in clinical decision making to date has been difficult to achieve, but the immersive nature of virtual reality now provides the necessary tools (Bremmer, Gibbs and Mitchell, 2019).

Research is working to prototype augmented reality applications in medicine such that a surgeon using an augmented reality headset would be able to see digital images and other data directly overlaid on the field of view. This has the potential to reduce intraoperative time, complications, and cost, ultimately leading to increased patient survival (Bremmer, Gibbs and Mitchell, 2019).

Off-shelf virtual reality technology using simple headsets has promise for the prevention and treatment of health conditions, particularly in psychological care. The immersive and potentially entertaining nature of virtual reality means that rehabilitation medicine has also been posited as an area which could benefit from virtual reality technologies. In prevention, virtual reality has been used by the military as part of predeployment training with the aim of reducing the incidence of post-traumatic stress disorder. The lifetime costs of post-traumatic stress disorders and if predeployment resilience training using virtual reality can reduce this, it is likely to be cost-effective (Bremmer, Gibbs and Mitchell, 2019).

Just as pertinently, while exposing those who have anxiety or post-traumatic disorders to powerful stimuli causing physiological response, therapists can be present alongside to guide, coach, and reassure patients in situ. The effect or realism with virtual reality allows a therapeutic window because scenarios can be repeated. The thought of facing a situation that triggers real physiological anxiety is never a pleasant one. The thought of facing it with support alongside, in what is recognised as a virtual scenario, is far more surmountable than it would be in real life (Bremmer, Gibbs and Mitchell, 2019).

The benefits of virtual reality can also be seen outside psychological therapy. In rehabilitation medicine, a study in patients' post-acute stroke, showed that the use of virtual reality in combination with conventional therapy improved outcomes compared with conventional therapy alone. Virtual reality has also been shown to enhance the enjoyment and intensity of physical activity, and by using omni-directional treadmills, potentially increased the activity of participants (Bremmer, Gibbs and Mitchell, 2019).

The big challenges in the field of virtual reality are developing better tracking systems, finding more natural ways to allow users to interact within a virtual environment and decreasing the time it takes to build virtual spaces. There are not many companies that are working on input devices specifically for virtual reality applications. Most virtual reality developers have to rely on and adapt technology originally meant for another discipline, and they have to hope that the company producing the technology stays in business. As for creating a virtual world, it can take a long time to create a convincing virtual environment – the more realistic the environment, the longer it takes to make it (Mandal, 2013).

1.1.2 Augmented Reality

FROM THE EVER-GROWING increase in medical complexity, there was born a need for technology to go beyond mere simulated reality. Augmented reality (AR) was the answer to this problem. AR is a data or information 'overlay' on the physical world that uses contextualised digital information to augment the user's real-world view (PwC, 2019b). Augmented reality presents digital information, objects, or media in the real world through a mobile device or headset. These elements can appear as a flat graphical overlay or can behave as a seemingly real '3D' object.

With AR, there is no created scenario; instead, an actual event is being altered in real-time, which has significantly enhanced robotic surgery. AR can work in parallel with a telemanipulation system in order to optimise the visual field of the operating surgeon. AR is currently being utilised to overlay key anatomic landmarks during live surgery to optimise patient safety (Moawad et al, 2020).

Impact on the world

Marketing and advertisement fields will explode with augmented reality devices. Augmented reality is the future that will allow consumers to experience a reality that is based on personal needs and desires. AR will present a completely new way to engage and will expand the abilities of retailers as well (PwC, 2019b).

Impact on education

AR provides the opportunity for virtual and immersive learning opportunities. AR allows different types of simulations which students can learn from – such as in aviation studies, where flying an aircraft can be simulated, or in medicine, where doctors can simulate an operation (PwC, 2019b).

Where this has been successfully used

- Eastern Michigan University: to help students grasp concepts in Earth science, a professor built an AR sandbox using a Microsoft Xbox Kinect camera, digital data projector, computer, and simulation and visualisation software. Students could create mountains, volcanoes, river channels, glacial deposits or virtual rain by manipulating a digital map projected on a box of sand.
- University of Canberra, the Australian National University and Macquarie University: The Special Collections using Augmented Reality to Enhance Learning and Teaching (SCARLET) project digitised rare books and manuscripts held in the John Ryland's library in Manchester. An app was created which enables students to see and handle original materials while providing an additional layer of imagery, resources and information to augment the learning experience.
- Stanford University and Ohio State University use Google Glass, and optical head-mounted device or 'pair of glasses', in medical education, which enables doctors to see an operation from a resident doctor's perspective and provide real-time guidance on a procedure (PwC, 2019b).

Chapter 2

Impact of Virtual and Augmented Reality on Healthcare and Medicine

2.1 Introduction

THE IMPACT OF VIRTUAL reality and augmented reality on healthcare and medicine encompasses the following categories: clinical care; medical education and training; patient safety and security; economic and financial aspects; organisational aspects, socio-cultural aspects; legal, regulatory and scalability aspects.

Immersive health technologies are revolutionising the delivery of frontline healthcare, therapeutic techniques, and research. They also offer great potential to improve the training of healthcare professionals through reality-simulation training (Bremmer, Gibbs and Mitchell, 2019).

Healthcare is already embracing VR and AR but so much more is possible. There is need for technology that creates a better experience for staff and patients, so these innovations need to be combined with the resources and cultural change required to improve productivity and, most importantly, patient outcomes (PwC, 2019).

The impact of virtual reality and augmented reality on the healthcare sector could be huge over the next decade, for front-line patient care and also for training. Potential boost of virtual reality and augmented reality in healthcare to GDP by 2030 is $350.9 billion. Virtual reality is already being used to give medical students greater access to operating theatres, where there are restrictions on the number of observers. The technology is also being used to enable consultants based in different locations to collaborate remotely and discuss upcoming surgical procedures. Virtual reality and augmented reality have clear benefits in healthcare. Being able to create operating theatres and realistic scenarios in virtual reality will help train doctors and surgeons and test their decision-making and responses to stressful situations in a risk-free way. Virtual reality can potentially be used therapeutically too, creating applications to help people cope with anxiety. Augmented reality glasses can overlay scans and x-rays onto a patient's body, augmenting the view a surgeon has. Similarly, augmented reality can help a doctor see at a glance a patient's test results and data, at the bedside, there and then, rather than logging into a desktop computer or checking paper notes. Healthcare is already embracing virtual reality and augmented reality but so much more is possible (PwC, 2019).

2.1.1 Clinical Care

SOCIETY AND HOSPITALS are evolving, and so must the ways we treat illness. Some experts believe that virtual reality (VR) and augmented reality (AR) can drive much of that change. VR can potentially be used therapeutically, creating applications to help people cope with anxiety. The technology is also being used to enable consultants based in different locations to collaborate remotely and discuss upcoming surgical procedures. AR glasses can overlay scans and x-rays onto a patient's body, augmenting the view a surgeon has. Similarly, AR can help a doctor see at a glance a patient's test results and data, at the bedside, there and then, rather than logging into a desktop computer or checking paper notes (PwC, 2019).

VR has been shown to support rehabilitation in clinical settings and to encourage at-home treatment adherence (Tran et al, 2021). The new generation of VR systems supports avatars (such as, cyberhands and soon wireless full body tracking, full avatar bodies, face tracking and eye tracking), and the addition of "embodied" presence is expected to enhance efficacy, as it enhances the patient's experience, increases patient's sense of "physical control" of their body (albeit a virtual body). Avatar VR may be useful for some chronic pain patients (Matamala-Gomez et al, 2019) and also for a wide range of acutely painful medical procedures (Hoffman, 2121).

Laparoscopy. VR simulators are widely used in laparoscopic training institutes, since acquiring specific surgical skills in the operating room can be inefficient, time-consuming, and potentially risky for the patients (Alaker et al, 2016; Yiannakopoulou et al, 2015).

Orthopaedics. Compared with laparoscopic and other disciplines, development and application of the VR simulators for orthopaedic surgery are lagging behind (Vaughan et al, 2016).

Other surgery. Andersen et al (2014) have reported an application of VR simulator in mastoidectomy performance assessment, using Visible Ear Simulator (VES) to complete this task. VES is a fully functional 3D virtual temporal bone simulator with force feedback.

According to Dr Shafi Ahmed, Advisor in Digital Health Transformation and Innovation, Department of Health Abu Dhabi, the value the VR and AR technologies bring to surgeries is not only cost reducing, but life-saving and accessible to everyone (PwC, 2019).

VR technology for acute pain management. Acute pain caused by operation or trauma induces a wide range of pathophysiological responses. Inflammatory, physiological and subsequent behavioural responses follow the activation of nociceptors through tissue injury. Sympathoneural and neuroendocrine activation, combined with uncontrolled pain, can ultimately lead to various detrimental responses (Carr and Goudas, 1999). The management of pain in the acute care setting often relies on pharmacological treatments, such as anaesthetic and analgesic agents, to attenuate these pathophysiological responses (Kehlet and Holte, 2001).

However, the arrival of VR technology brought another path for the management of acute pain. Several studies have confirmed that standard analgesia coupled with VR based games is effective in reducing acute pain (Hoffman et al, 2001; Sharar et al, 2007; Carrougher et al, 2009). Recent reports highlight the safety and efficacy of VR distraction for the management of acute pain during painful medical procedures (Le May et al, 2021; Hoffman et al, 2019).

Another cutting-edge approach for use of VR technology in pain management refers to augmenting hypnosis, also known as virtual reality hypnosis (VRH). The use of VR for relaxation, in addition to use of morphine for pain reduction have shown to be very effective (Konstantatos et al, 2009).

VR technology for chronic pain management. Chronic pain is typically defined as pain lasting longer than three months, or beyond the expected period of healing (Saravanakumar, 2010). Some types of chronic pain typically develop in an arm, hand, leg or foot, or lower back after an acute trauma, and are characterised by persistent pain, and in some cases, greatly exaggerated and debilitating sensitivity to pain. Chronic pain conditions are difficult to treat because of non-effective traditional pharmacologic treatments and invasive nerve blocks.

Although the pathogenesis of chronic pain is still poorly understood, central sensitisation, by which nociceptive input can trigger increased excitability of neurons in central nociceptive pathways, is likely a relevant mechanism in some types of chronic pain (Bruehl and Warner, 2010; Woolf, 2011).

It is not unusual for the brains of some amputees to register pains coming from their severed extremities. Sweden's Ture Johanson, for instance, has suffered from phantom limb pain (PLP) since losing the lower part of his right arm in a car accident more than 50 years ago. He tried all the usual therapies, but none worked. Then he became the initial patient to try a radical new AR treatment developed by Max Ortiz Catalan, a doctoral student in biomedical engineering at Sweden's Chalmers University of Technology (First Look, 2014).

Although very little is known about using AR for pain distraction, the use of see-through augmented reality goggles may prove valuable on the battlefield or during Medivac (Petterson et al, 2021). In addition, calm, non-nauseogenic VR worlds such as SnowWorld (Hoffman et al, 2019), are specifically designed to minimise motion sickness.

Recent advances in functional imaging technology have revealed that chronic pain conditions are caused by neuronal plasticity in the central nervous system (CNS), even though the underlying pathogenesis has not been fully understood (May, 2008). Only a few studies have investigated the use of VR for chronic pain management, therefore very little is known about the use of VR for chronic pain treatment and for long-term rehabilitation (Li et al, 2011). Some of the studies have demonstrated that VR system might increase analgesic efficacy and provide a promising alternative treatment for complex regional pain (Sato et al, 2010).

VR can affect pain perception through immersive virtual environment, by occupying finite attentional resources and by blocking external stimulation associated with real environment and the painful stimuli (Wismeijer and Vingerhoets, 2005). Since distraction interventions work by competing for attention otherwise directed towards painful stimuli, pain tolerance and pain threshold have shown to increase under VR conditions (Hoffman et al, 2001; Rutter, Dahlquist and Weiss, 2009). Pain intensity, anxiety and time spent thinking about pain have shown to decrease following VR distraction (Weiss, 2009). Immersive VR technology is more likely to generate relief from pain compared to non-immersive VR technology (Gold et al, 2006).

Although research results support the premise that VR devices can reduce pain, the neurobiological mechanisms still need to be determined. The current state of VR as a tool for pain management is still in its early developmental stages and it requires numerous applications for patients with an array of acute and chronic medical conditions. Nonetheless, VR technology will eventually emerge as a promising first-line intervention and complementary therapy for patients with pain (Li et al, 2011).

In addition to its application for acute and chronic pain management, VR is increasingly being used for the treatment of the frequently accompanying psychiatric challenges that are commonplace such as anxiety and depression. Therapeutic VR is showing promise for the treatment of these common occurring psychiatric symptoms (Petterson et al, 2021).

Emerging reports point to the readiness of clinicians to add VR to their highly effective telehealth and telepsychology practice, particularly during the global COVID-19 pandemic (Sampaio et al, 2021).

VR technology for psychological diseases therapy. The treatment for many psychological disorders commonly require patients to confront the situations they fear. This kind of approach, also known as exposure therapy, helps patients to accept their anxious emotions and consequently change their beliefs about the likelihood or catastrophic nature of feared consequences. Exposure therapy is very effective but hard to execute, since complex conditions such as social situations, thunder, injury and other phobic stimuli are potentially very challenging to recreate and to expose individuals to the same in a real life setting (Valmaggia et al, 2016). Consequently, virtual environment created by VR simulations could be a valuable option for exposure therapy (Gega, 2017).

VR devices can be used in private practices and even private homes, which is very important since it may additionally aid patients to cope with unwanted feelings in "emotionally safe environments" (Li et al, 2011). According to existing publications, the software is the core part affecting the treatment outcomes. Virtual models and environments designed for medical therapies vary according to the patient's specific symptoms. Every patient's symptom is different, each VR platform has diverse working conditions and therapy procedures largely depend on individual therapist's decisions. An appropriate evaluation system for

VR therapy alone, or for VR therapy combined with other therapeutic methods are urgently needed. Treatment of psychological diseases is very time-consuming, thus, VR devices providing plentiful and vivid contents may be of great help for patients going through difficult periods (Li et al, 2017).

New technologies such as VR and mobile applications may stimulate help-seeking and reduction of stigma about psychological conditions such as post-traumatic stress disorder (PTSD). These technologies can be used for education, initial screening of PTSD symptoms and assignation to treatment. The delaying factors for militaries in help seeking underscore the vital importance of adequate screening and detection of PTSD to facilitate timely treatment and prevent symptoms from worsening (van Bennekom and de Koning, 2018).

"Advances in simulation and gaming technology have resulted in the ability to create emotionally responsive, three-dimensional virtual humans (VHs) that possess personality, memory, and non-verbal gestures, and react like real people engaged in health conversations. As a result, individuals experiencing signs of PTSD, substance use, and suicide risk often report feeling less judged, safer, and more likely to reveal information to VHs when compared to face-to-face conversations of a similar nature (Lucas et al, 2014; Rizzo et al, 2016; Pickard, Roster and Chen, 2016). This is due to the development of algorithms that enable VHs to consistently and reliably establish rapport with users, provide accurate knowledge, and can react to users, provide accurate knowledge, and can react to user responses with evidence-based communication strategies such as motivational interviewing (MI) (), all contributing to their effectiveness and high fidelity. Thus, within conversation dynamics, VHs can respond predictably and efficaciously, never fatigue, do not age, and are not subject to transference and countertransference reactions that can compromise communication such as a provider's own sets of expectations, beliefs, biases, and emotions that may impact the quality of care. Clearly, VHs hold tremendous potential to leverage healthcare conversations resulting in changing attitudes and behaviours" (Albright and McMillan, 2018).

2.1.2 Medical Education and Training

ADVANCEMENTS IN TECHNOLOGY continue to transform the landscape of medical education. There has been a recent dramatic expansion in the ways in which we can deliver medical education. The need for technology-enhanced distance learning has been further accelerated by the coronavirus disease (COVID-19). This has not only been through the internet and mobile devices, but through immersive technologies. Emerging technologies, including virtual reality (VR), augmented reality (AR), and alternate reality present unique approaches to teaching clinical skills to all levels of learners using computer-generated simulations. Computer-generated simulation uses a human-computer interface to experience and interact with and varies in terms of immersion, fidelity, and interactivity (Kassutto, Baston and Clancy, 2021; Pottle, 2019).

VR in particular has been adopted across medical and nursing fields. VR involves the user putting on a VR headset to become completely immersed in an interactive virtual environment. When used with appropriate educational software, this allows the user to learn from experience in the virtual world (Pottle, 2019)

In medicine, staff can use virtual environments to train in everything from surgical procedures to diagnosing a patient. Surgeons have used virtual reality technology to not only train and educate, but also to perform surgery remotely by using robotic devices. Being able to create operating theatres and realistic scenarios in VR will help train doctors and surgeons and test their decision-making and responses to stressful situations in a risk-free way. If further progress is made in this field of virtual reality in healthcare, it has the ability to revolutionise (Janani, Arthy and Somasundaram, 2011).

For decades, the acquisition of technical skills in the operating room under the supervision of senior surgeons has been the only way for junior doctors to receive surgical training (Aim et al, 2016). As the number of trainees has increased, the opportunities to acquire necessary technical skills have become increasingly limited due to the higher costs, ethical concerns, and decreasing resident work hours. As the surgical techniques have advanced and evolved, sole observation was no longer enough for acquiring certain skills and special techniques. In these circumstances, VR training has become an essential prerequisite for junior doctors before they are allowed to actively participate in real operations.

Compared with animal modes, videos and e-learning, VR simulations are more realistic due to very intuitive anatomic structures exhibited in 3D. Trainees can interact with all the anatomical structures, including skin, muscle, bone, nerve and blood vessel. Changes that occur following each surgical step are very much the same as in reality. Whole performance can be recorded, compared and analysed, making data permanently available for trainees (Aim et al, 2016). From different a perspective, senior supervision and patient participation are no longer needed during the period of basic skill training and acquisition, since the VR simulations can provide a controlled virtual environment necessary to satisfy these requirements outside the operating room (Yiannakopoulou et al, 2015).

As an upcoming tool in surgery training programmes, VR simulators focus on the advanced stages of training. They can provide automated scoring with numerous objective metrics that are very promising alternative for the laborious and subjective ratings usually performed by experts during live or videotaped procedures (Van Bruwaene et al, 2014; Berg et al, 2007; Vassiliou et al, 2007).

Haptic experience in surgical simulation can help trainees to familiarise themselves with the operation process and to develop their operative and decision-making skills without potential harm to patients (Tonetti et al, 2009; Petterson et al, 2008). However, the accuracy and efficacy of VR training still need to be improved. Density, palpable properties and convex surfaces are very challenging to simulate in a virtual environment. Moreover, the response rate and intensity feedback or commercial haptic devices are too weak to correctly emulate vibration while drilling (Mark et al, 1998).

An ideal VR simulator for surgical skills training must satisfy several conditions, including multimodal training plans, artificial platform, accurate haptic feedback, immersive visual and audio technologies, sensitive input devices, as well as appropriate software with real-time simulation and assessment criteria. In addition, it should incorporate data from MRI or CT to provide patient-specific simulations. Combination of physical body models promotes the interaction between realistic tools, virtual bodily fluids and virtual physical materials. Abnormal and patient-specific cases can be modelled and assessed in VR simulators for trainees to practise and to prepare in vivo. These features and limitations may provide some directions for the future development of VR technology for surgical skills training (Vaughan et al, 2016).

Despite the advantages of VR in medical education and training, VR simulation is not a panacea. Rather, it is a tool used to accomplish a defined set of learning outcomes and should be deployed as such, integrated within an institution's curriculum and pedagogy to ensure effective use. For example, VR is not suitable for every possible educational opportunity. It is not the best way to teach abdominal palpation; there is no need for complex immersion in this situation, just an accurate physical representation of an abdomen (Pottle, 2019).

Virtual characters are often controlled by artificial intelligence (AI) systems. Though this is developing fast, it is not yet suitable for certain learning objectives, such as breaking bad news. The complexities of language processing and facial expression are, at present, best covered by a human rather than a virtual patient (Pottle, 2019).

VR technology applications in medical education and treatment, especially psychiatric treatments need a comprehensive manual that specifies how, where, and for whom this technology is appropriate. Individual characteristics (gender, age, personality, and history of motion sickness) and other unique psychological, cognitive, physical, and functional characteristics that are common in a variety of clinical conditions are very important and should be considered. Sensitivity in some patients need to be considered, including apprehensiveness to use head-mounted display (HMD), the reality test itself, learning capacity to act in a virtual environment, and the potential side-effects and duration. These issues must be specified, because of ethical reasons and the effectiveness of education and treatments based on VR (Nichols and Patel, 2002; Rizzo, Strickland and Bouchard, 2004).

Successful implementation of modern technology like VR depends on scheduling suitable educational training. In other words, the training of end-users in launching any new system is one of the success factors in the implementation and execution of programmes (Lluch, 2011).

Studies show that patient age is a key factor in treatment appropriateness when using VR. For example, VR-based therapy that is suitable for children's treatment may not be useful for older people and may provide different outcomes. Use of VR is not appropriate for the treatment of severe anxieties and may have negative effects. VR applications for people that are suffering from a particular type of psychiatric illness or different psychiatric properties must be done more carefully (Botella et al, 2009). Insufficient training of clinicians in the proper use of this technology for therapeutic purposes will have adverse consequences.

Besides education, preparing users according to different dimensions is very important. Neglecting user readiness confronts VR with a lot of challenges (Rizzo, Strickland and Bouchard, 2004; Rizzo, Schultheis and Rothbaum, 2011).

2.1.3 Patient Safety and Security

STUDIES HAVE GENERALLY reported VR can be safely used in children in a healthcare setting, and that few or no side effects are associated with VR for paediatric patients (Caruso et al, 2020; Gold et al, 2006; Morris, Louw and Grimmer-Somers et al, 2009; Gold and Mahrer, 2018; Faber et al, 2013). Adverse events, although rare and short-lived, may include motion sickness, nausea, or dizziness.

The risk of collision also exists for specific cases of VR use. Additionally, paediatric patients may be less able to verbalise VR-associated discomfort, and young children may require contextualisation after interacting with a virtual world. As VR becomes more widely used in paediatric hospitals, it is critical to understand its safety and efficacy profile in this setting (Caruso et al, 2020).

Baniasadi et al (2020) caution that "like any other type of treatment, VR should only be used when it is prescribed by the appropriate clinical expert. Using services such as VR rehabilitation or any type of emerging technology such as telepsychology and online therapy, according to the patients' opinion (self-diagnosis, self-help, and self-treatment) can place patients at potential risks (Rizzo, Strickland and Bouchard, 2004; Rizzo et al, 2002; Botella et al, 2009).

VR can cause problems in the cognitive organisations, human experiences, memories, judgements, beliefs, and distinguishing between themselves and the environment. Having multiple virtual experiences over the years can make real judgement and self-identity difficult (Botella et al, 2009). Therefore, managing the health and safety implications of VR are important things to be considered (Nichols and Patel, 2002). Regarding telerehabilitation, home-based VR therapy without the direct supervision of the therapist may pose certain risks (Rizzo, Strickland and Bouchard, 2004).

Most of the people may experience positive aspects of using VR, such as improvement in visualisation performance or pleasure performance more than any negative effect. With the advance of technology, some of these effects may be reduced, in contrast, some may remain or even become worse. Therefore, a prospective approach is required in order to allow empirical studies to be continued and ensures that experimental data can be converted as guidance (Nichols and Patel, 2002).

Cybersickness and perceptuomotor after-effects have been reported as the potential side- effects of VR. In fact, the main concern for VR users is simulation sickness, or cybersickness (Srivastava and Das, 2014; Bohil, Alicea and Biocca, 2011; Rizzo, Strickland and Bouchard, 2004; Nichols and Patel, 2002). Also, headaches and eye strain are seen in prolonged exposure with VR systems (Garrett et al, 2018). Similar to motion sickness, cybersickness (for example, nausea, vomiting, among others) is related to the conflict between different body sensory systems, which is due to an inconsistency between the sensory inputs. Some cybersickness cases persist such as when the real gravity is in contrast with the seen environment, for example, when the user is flying in an airplane in VR (Bohil, Alicea and Bocca, 2011).

Additionally, VR users face technological limitations in manipulating tools and components in the virtual world resulting in distraction in learning due to interference with VR (Adapa et al, 2020). On the other hand, AR removes these limitations in learning by blending into the user's reality with minimal interference; that is, the users do not completely immerse into virtual space and can see the superimposed objects within the real-world environment. This enables the AR users to be more interactive with the task in hand (Adapa et al, 2020).

From an ethical perspective, patients, and therapists should be aware of the potential risks of technologies such as VR. This awareness should result in making rational decisions for its safe and effective use in the early stages of development of the programme (Rizzo, Schultheis and Rothbaum, 2002; Lockwood, 2004).

2.1.4 Economic and Financial Aspects

ALMOST EVERY INDUSTRY is responding to a growing need to deliver greater innovation, faster and more effectively, at better prices. That means finding a way to develop products more quickly and more efficiently, reducing time to market while taking cost and complexity out of the process. This is where VR and AR can make a huge difference. They can transform the way businesses develop products. Rapid prototyping and collaboration using VR and AR technology will speed up and enrich the creative process. The time-intensive requirement to build physical prototypes can be reduced dramatically, bringing ideas and innovations to life and products to market far more quickly. But these technologies are not just a creative means to an end. They also offer the potential to create new services and engage consumers in unprecedented ways. The impact of VR and AR on the healthcare sector could be huge, and it has a potential boost of $350.9 billion to GDP by 2030 (PwC, 2019).

Previous research stated that the implementation of VR is expensive (Srivastava, Das and Chaudhury, 2014; Bohil, Alicea and Biocca, 2011; Rizzio, Strickland and Bouchard, 2004; Botella et al, 2009). This is one of the obstacles to be met before the extensive application of the VR (Srivastava, Das and Chaudhury, 2014). Establishing and developing VR programmes requires high-quality hardware, high-speed computers, efficient graphic cards, accurate tracking systems, high-resolution displays, and highly specialised accessories (Rizzio, Strickland and Bouchard, 2004). Initial VR tools have problems, including their large size and lumpiness, the difficulty of using them, and the high cost of design and implementation, which slow down the process of using these tools (Bohil, Alicia and Bioca, 2011).

Design content of VR depends on the involvement and cooperation of different professions in various fields which can make it too expensive (Nunes and Costa, 2008).

There will be benefits and shortcomings of implementing virtual reality in healthcare and medicine for each country depending on whether the country has a national systems-based policy, market-driven policy or insurance driven policy. The shortcomings can be addressed by adopting a comprehensive systems-based policy to ensure that there is a win-win situation where all players or stakeholders reap the benefits of virtual reality in healthcare and medicine.

Virtual patients are seen as cost-effective since virtual patients are limiting the effort and expense associated with standardised patient training and can be delivered at low cost over the internet. Concerning the challenges of developing and using virtual patients, mainly legal and technical challenges were mentioned. Legal issues included management of rights, permissions, and copyright issues. Technical issues were, for example, management of hard- and software, technology support, unreliable internet connection, difficulty in editing a virtual patient, or cross-platform compatibility. Other challenges were low content validity and reliability, difficulty of integration into a curriculum, and a non-realistic, impersonal, and isolated learning experience (Hege et al, 2016).

2.1.5 Organisational Aspects

THERE ARE STILL QUESTIONS and uncertainties about how other communication technologies connect to virtual worlds and how websites will connect with virtual worlds, bearing in mind the heterogeneity of virtual worlds in healthcare and medicine. It is still not clear how software developers can design an easy-to-use and powerful application to build content for virtual healthcare and how individuals can run their own virtual world servers and interact with other serves in healthcare and medicine and beyond (Janani, Arthy and Somasundaram, 2011).

There are several "communities" using the term virtual patient for their activities. Within these communities it might be obvious what is meant by the term, but when communities and exchanging research results between such communities and to non-virtual patient experts a clear understanding is indispensable. The primary form of virtual patients in the educational literature are Interactive Patient Scenarios despite rapid technical advances that would nowadays support more complex applications. The adapted classification provides a valuable model for virtual patient developers, educators and researchers in healthcare education to more clearly communicate about the virtual patients they are using (Konowicz et al, 2015).

In countries such as Iran, VR approaches have been less applied in medicine, especially in the field of education and treatment. Lack of clinicians' knowledge and awareness about this technology is one of the contributing factors. Identifying technical and non-technical limitations have a pivotal role in the successful and proper use of these systems. Lack of suitable standards, insufficient infrastructure, difficulties in content providing, organisational culture, and management support are some of these limitations. In general, planning for the study of experiences of the leading countries in the use of these technologies, developing and updating related laws, guidelines, standards, and utilising appropriate models in the design and implementation of VR-based applications should help in using VR in medical education and treatment. Also, there is a need to determine appropriate policies that will play an important role in the successful implementation of these technologies at the national level (Baniasadi et al, 2020).

Developing Virtual Healthcare Systems in Complex Multi-Agency Service Settings: the OLDES Project. The OLDES project aims to offer new technological solutions to improve the quality of life of older people. This intervention is intended to facilitate the development of a system that more purposively fits the everyday practices and needs of older people and the agencies and institutions that provide care services for them. This activity is supported by the deployment of 'demonstrator' tool that supports stakeholders in visualising different socio-technical scenarios for the configuration of virtual services. There is potential for this tool to support processes of shared sense making amongst multiple care agencies and institutions in complex care environments. It is believed that this activity will result in an OLDES system platform for a virtual healthcare system that is more likely to be valued by those whose needs it is intended to serve and those who exercise a duty of care in meeting those needs (Maniatopoulos et al, 2009).

Evaluation and validation of VR applications. Science-based standards development with the collaboration of specialised teams is essential for evaluating VR systems. The continuity of evaluation, as well as the use of a combination of quantitative and qualitative methods for evaluations, especially usability evaluation, should also be considered outside the laboratory and in the clinical setting. Also, the responsibility of programme producers and project team should be defined in the form of guidelines and rules to specify who will be responsible in the case of a problem occurring for the users by the use of these modern tools. VR applications should not be used before validating programmes and providing the operating instructions (Baniasadi et al, 2020).

The acceptance of VR and modern technologies by patients and therapists as well as students for educational use should be considered an essential principle. Managing and monitoring the health of users of VR programmes is necessary. Moreover, diagnostic and preventive plans for the change in individuals' behaviour after applying VR-based games and applications should be considered. It should be emphasised that these tools are used as complementary to the usual methods, and not a substitute for it (Baniasadi et al, 2020).

Purpose-based and standard design of VR-based educational and therapeutic programmes requires collaboration between medical and IT experts as well as end-users by their feedback and comments to provide the effective content and successful implementation of the programmes.

Using a UCD approach to design usable programmes is recommended. In programmes with a focus on education, students need to be aware of the limitations of interactive electronic content, such as failing to provide some details in designing programmes (Baniasadi et al, 2020).

Creating appropriate strategic guidelines based on many practical studies and their results are essential in the use of any type of new technology such as VR.

2.1.6 Socio-cultural Aspects

THERE IS POTENTIAL for virtual healthcare to address inequalities in healthcare in the contexts of both developing and developed countries. In countries such as New Zealand, there is potential for virtual healthcare to bridge the gap for marginalised and vulnerable populations living in remote settings by increasing healthcare access for children and families who may otherwise face significant barriers in having their health needs met (Field and Butler, 2018). There are barriers, though, that need to be taken into consideration in the implementation of virtual healthcare such as slow adoption of such technology by older patients and the fact that those identified to benefit from such technology could be the most likely not be able to pay for such services because they cannot afford (Field and Butler, 2018). There is also the possibility that virtual healthcare could not be accepted by both health professionals and/or patients.

Given that there are many standards, there is uncertainty about what the main three-dimensional protocol for Virtual Worlds in healthcare and medicine will be. The ethics related to virtual marketing in healthcare and medicine have not been defined at national, regional and global level. It has also not been clearly defined how consumers will be protected from some kinds of virtual marketing in healthcare and medicine (Janani, Arthy and Somasundaram, 2011).

Users' attitude and consequently the modern technology's acceptance by the users will be the key non-technological factor in the utilisation of technological tools by individuals (Cresswell and Sheikh, 2013; Mohammadzadeh and Safdari, 2014; Venkatesh, Thong and Xu, 2012). For example, some patients may be reluctant to use computers and such new technologies, which should be considered before starting treatment (Rizzo, Strickland and Bouchard, 2004). Moreover, people who are unfamiliar with information technology (IT) may also resist the use of new technologies and cannot trust to use IT-based tools, especially if they do not have the required IT skills (Garrett et al, 2018; Botella et al, 2009; Al-Mujaini et al, 2011).

According to Alex Foden, Director, Change Communications and Culture, PwC UK, "VR and AR offer incredible benefits, but those benefits will not be realised if organisations do not bring their people along with them. If technology fails to deliver on its promise, it is often because an implementation was planned without people in mind or a plan to engage them. Start with the people and communicate clearly what the ambition is, what the benefits will be made and what the business case is. With technologies such as VR and AR, create VR and AR experiences that not only convey the benefits, but consider how they can actually be used as part of the communication process" (PwC, 2019).

2.1.7 Legal, Regulatory and Scalability Aspects

LEGAL ISSUE. CAN GOVERNMENT offer the real-world services to real-world citizens and businesses more efficiently in Social Virtual Worlds? Will e-government evolve to v-government? How will our virtual social interactions impact our real social behaviour? How will real life laws and rules apply to virtual worlds? Is it necessary to define special laws and rules for virtual worlds? What are the users' responsibilities when committing crimes through their avatars? What is the ethics related to virtual marketing? How will consumers be protected from some kinds of virtual marketing (Janani, Arthy and Somasundaram, 2011)?

2.2 Conclusion

VIRTUAL REALITY TECHNOLOGY is widely applied in the field of medicine. Huge benefits have been reported following its use for rehabilitation, disability management, surgical training, psychological diseases therapy and analgesic modalities (Baldominos, Saez and Del Pozo, 2015).

VR is already transforming medical education. VR offers distinct benefits for learners, faculty and the health system. It is helping to free learning from the classroom, allowing learners to apply their knowledge to practise and learn from mistakes. It focuses on improving competencies and places the emphasis on autonomous, blended learning, which is expected from the learners of today (Pottle, 2019).

As well as financial savings, such technologies free up space and faculty time. Faculty do not necessarily need extra training to be able to use the VR equipment. From a global health perspective, this reduction in cost and equity of access allows simulation to be distributed globally. This potential to democratise the availability of quality medical training makes VR an exciting prospect in healthcare training. As VR continues to be implemented and integrated within curricula, its use will become mainstream. The ability for multiple learners to take part in truly inter-professional, completely life-like simulation which is not bound by geography, is set to change how we conduct medical and inter-professional education beyond recognition (Pottle, 2019).

Virtual patients are seen as cost-effective since virtual patients are limiting the effort and expense associated with standardised patient training and can be delivered at low cost over the Internet. Concerning the challenges of developing and using virtual patients, mainly legal and technical challenges are mentioned. Legal issues include management of rights, permissions, and copyright issues. Technical issues are, for example, management of hard- and software, technology support, unreliable internet connection, difficulty in editing a virtual patient, or cross-platform compatibility. Other challenges are low content validity and reliability, difficulty of integration into a curriculum, and a non-realistic, impersonal, and isolated learning experience (Hege et al, 2016).

The design of a secure and trustful community is a difficult though interesting task, which should be preferably performed by following standardised procedures. Security in healthcare communities is not a simple task, and there is not a single solution or remedy. The roadmap for developing a secure solution will soon deteriorate if the suggested risk management model is not iteratively repeated. With well-defined procedures and workflows, this standardised methodology achieves a secure environment for the healthcare community and allows medical professionals and patients to act inside the community with minimum risk for their valuable assets (Chryssanthou, Varamis and Latsiou, 2011).

There are several "communities" using the term virtual patient for their activities. Within these communities it might be obvious what is meant by the term, but when communities and exchanging research results between such communities and to non-virtual patient experts a clear understanding is indispensable. The primary form of virtual patients in the educational literature are Interactive Patient Scenarios despite rapid technical advances that would nowadays support more complex applications. The adapted classification provides a valuable model for virtual patient developers, educators and researchers in healthcare education to more clearly communicate about the virtual patients they are using (Konowicz et al, 2015).

Virtual reality brings new possibilities and facilitates healthcare with a positive experience for the treatment of the patient by creating a virtual 3D environment. It gives a better perception of the surrounding environment. This technology has diverse applications in neuroscience, psychology, physical and occupational therapy and other intervention approaches and the potential for the treatment of stress-related disorders. In the medical field, this technological revolution seems the right solution, which can allow immersing the situations. It can reduce therapist consultation duration. This technology is now useful for taking up unique patient cases and challenges. For mental health issues such as dementia, depression and stress management, this technology provides an effective and better solution which enhances patient and creates a positive effect to save the life of the patient (Javaid and Haleem, 2020).

In medicine, staff can use virtual environments to train in everything from surgical procedures to diagnosing a patient. Surgeons have used virtual reality technology to not only train and educate, but also to perform surgery remotely by using robotic devices. If further progress is made in this field of virtual reality in healthcare, it has the ability to revolutionise (Janani, Arthy and Somasundaram, 2011).

The introduction of patients and clinicians to virtual environments raises particular safety and ethical issues. Despite developments in virtual reality technology, some users still experience health and safety problems associated with virtual reality use. It is however true that for a large proportion of virtual reality users, these effects are mild and substitute quickly. The temporary side effects can be divided into three classes of symptoms related to the sensory conflicts and to the use of virtual reality equipment: visual symptoms (eyestrains, blurred vision, headaches); disorientation (vertigo, imbalance) and nausea (Claudio and Maddalena, 2014).

Despite the benefits of using virtual reality in medical education and treatment, some challenges and limitations result in the uselessness or misuse of this technology. Therefore, recognising potential challenges related to virtual reality might be helpful in the strategic decision-making process to implement and develop this technology in the healthcare and medical field (Banlasadi, Ayyoubzadeh and Mohammadzadeh, 2020).

The clinical applications, behavioural aspects, and technological developments in virtual reality and augmented reality research are parts of a more complex situation compared to the old platform used before the huge diffusion of head mounted displays (HMDs) and solutions (Cipresso et al, 2018).

Chapter 3
Selected Studies

3.1 Study 1: Virtual Painkiller

FINDINGS AND CONTRIBUTIONS

"It is not unusual for the brains of some amputees to register pains coming from their severed extremities. Sweden's Patient X, for instance, has suffered from phantom limb pain (PLP) since losing the lower part of his right arm in a car accident more than fifty years ago. He tried all the usual therapies, but none worked. Then he became the initial patient to try a radical new augmented-reality treatment developed by Max Ortiz Catalan, a doctoral student in biomedical engineering at Sweden's Chalmers University of Technology. Patient X's stump was wired to a computer that picked up the electrical impulse in the brain coming from the severed extremity. The augmented reality treatment worked in relieving Patient X from the phantom limb pain" (ASEE Prism).

3.2 Study 2: Virtual World Technology for Healthcare: A Survey

FINDINGS AND CONTRIBUTIONS

In medicine, staff can use virtual environments to train in everything from surgical procedures to diagnosing a patient. Surgeons have used virtual reality technology to not only train and educate, but also to perform surgery remotely by using robotic devices. If further progress is made in this field of virtual reality in healthcare, it has the ability to revolutionise (Janani, Arthy and Somasundaram, 2011).

Challenges facing Virtual World implementation can be divided into technical and social and legal issues. Technical issues: Given that there are many standards, what will be the main three-dimensional protocol for Virtual Worlds? How will other communication technologies, such as personal digital assistant and mobile phones connect to virtual worlds. How can software developers design an easy-to-use and powerful application to build content for virtual worlds? How will websites connect with virtual worlds? How will individuals run their own virtual world servers and interact with other servers? How will heterogeneous virtual worlds be connected? How will v-business applications be implemented in virtual worlds (Janani, Arthy and Somasundaram, 2011)?

Social and legal issues: Can government offer the real-world services to real-world citizens and businesses more efficiently in Social Virtual Worlds? Will e-government evolve to v-government? How will our virtual social interactions impact our real social behaviour? How will real life laws and rules apply to virtual worlds? Is it necessary to define special laws and rules for virtual worlds? What are the users' responsibilities when committing crimes through their avatars? What is the ethics related to virtual marketing? How will consumers be protected from some kinds of virtual marketing (Janani, Arthy and Somasundaram, 2011)?

3.3 Study 3: Virtual reality, augmented reality... I call it i-Reality

FINDINGS AND CONTRIBUTIONS

Futurist and Clinician, Dr Rafael Grossmann of the Eastern Maine Medical Centre, Bangor, United States of America, has concerns about the worldwide physician deficits as reported by the World Health Organization. He realises that new legislation and proposals will only add more patients to the system with a decrease in the number of medical students and deficit of physicians to take care of the patients. At the same time, Dr Grossmann realises the fact that a large proportion of the provider-patient interactions do not require physical presence. As such, he considers that virtual and augmented reality could improve the health system in two vectors:

- From the patient to provider direction: "Avatars" of providers that could interact with the patient and at least get from them most of the information that does not require "touch"; mainly medical history details);
- From the provider to the patient direction: The healthcare delivery person wearing a headset mounted device that would "read/detect" and create an estimated version of the patient, with medical data streamed by wearables on the patient and whose input is translated, even analysed by the provider's device.

Study 4: Reducing the stigma on posttraumatic stress disorder in militaries through virtual reality

FINDINGS AND CONTRIBUTIONS

Van Bennekom and de Koning (2018) report that in clinical research, the stigma and underreporting of posttraumatic stress disorder (PTSD) in militaries has been an area of investigation for decades. The delaying factors for militaries in help seeking underscore the vital importance of adequate screening and detection of PTSD to facilitate timely treatment and prevent symptoms from worsening.

New technologies such as virtual reality and mobile applications may stimulate help-seeking and reduction of stigma. These technologies can be used for education, initial screening of PTSD symptoms and assignation to treatment.

In a study by Lucas et al (2014), US active service militaries were interviewed by virtual interviewers about PTSF symptoms. The virtual human interviewers were created using the MultiSense system. This system is able to detect and analyse behavioural signals of the militaries such as head position, facial expression, body posture, and speech dynamics. The virtual interviewer can then provide an appropriate response such as a follow up question, an acknowledging gesture or feedback.

The study found that active duty militaries were more likely to disclose PTSD symptoms to the virtual interviewer in comparison to both the regular and anonymised version of the Post Deployment Health Assessment (PDHA). This underscores the potential of virtual reality in the initial screening for PTSD symptoms in militaries. This could also be beneficial to militaries who have limited access to mental health care facilities. The use of intelligent virtual therapies in combination with exposure treatment using multimedia including virtual environments could potentially facilitate an anonymous form of therapy.

A good mental condition of militaries benefits all parties including patients, their families, their employers and the people of the countries they are protecting (van Bennekom and Koning, 2018).

The study conducted by Lucas and colleagues has made a significant contribution in our understanding of, and the potential for using, Virtual Humans in conducting screenings that utilised the PDHA symptom checklist with active duty service members and veterans.

Analysis and Discussion

THE ANALYSIS AND DISCUSSION of the selected case studies highlights the strengths, weaknesses, opportunities and threats in the context of the impact of virtual reality on healthcare and medicine in developing and developed countries. In that regard, Figure 1 summarises the strengths, weaknesses, opportunities and threats of the impact of virtual reality on healthcare and medicine in developing and developed countries. Table 1 is a scanning of the environment using the SWOT and political, economic, social, technological, legal, and environmental (PESTLE) analysis. The analysis and discussion is broken down into the following sub-topics: impact on cultural and social conditions in healthcare and medicine, impact on regulatory conditions in healthcare and medicine, impact on knowledge and technical conditions in healthcare and medicine, impact on financial and market conditions in healthcare and medicine.

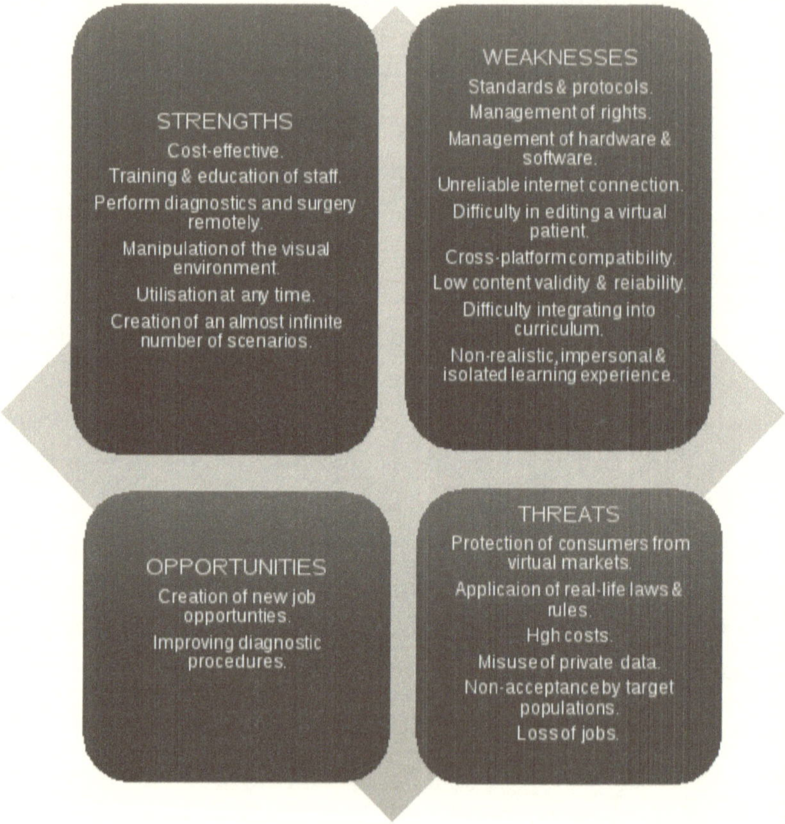

FIGURE 1: SWOT analysis of virtual reality in healthcare and medicine

Table 1: Scanning the environment of virtual reality in healthcare and medicine using the SWOT and PESTLE analysis

	S	W	O	T
P	Global political focus and pressure on healthcare and medicine.	Britain voted to leave European Union causes political turmoil.	Desire to make better use of existing public infrastructure investments, without requiring major new funding.	Political instability in some countries and regions.
E	Cost saves to countries due to reduced disease incidence rates.	Consumer confidence is low.	Market is likely to grow due to aging population.	Global economic crisis.
S	Patient awareness changing expectations.	Ethical concerns – replacing jobs in the long run.	Need for education and more price transparency.	Increasing age of population and growth in NCDs.
T	Advances in virtual reality can greatly increase the effectiveness of Healthcare delivery.	Interoperability issues.	New digital opportunities	Lack of infrastructure in many developing country settings.
L	Fines for organisations or institutions that do not comply with standards.	Legal compliance issues.	Need to focus on education and quality standards.	Unable to rationalise (different markets require different formulas).
E	Growing environmental agenda and community awareness	Power consumption demands.	Eco opportunities to market.	Adverse weather conditions cause temporary suspension of some operations.

Chapter 4
Future Perspectives and Critical Reflections

"*Everything is in place for virtual and augmented reality technologies to now deliver on their promise by improving the way the healthcare and medical ecosystem operates, making processes faster and more effective, creating incredible new experiences.*" – **Jeremy Dalton, Head of VR/AR, PwC United Kingdom**

COMPUTATIONAL METHODOLOGIES and modelling play a growing role for investigating mechanisms, and for the diagnosis and therapy of human diseases. This progress gave rise to computational medicine, an interdisciplinary field at the interface of computer science and medicine. The main focus of computational medicine lies in the development of data analysis methods and mathematical modelling as well as computational simulation techniques specifically addressing medical problems (Zlatko, 2012).

Society and hospitals are evolving, and so must the ways we treat illness. VR/AR can drive much of that change. The impact of VR and AR on the healthcare sector could be huge over the next ten years, for front-line patient care and also for training. It is estimated that the impact of VR/AR on the healthcare sector will have a potential boost of $350.9 billion to GDP by 2030.

Being able to create operating theatres and realistic scenarios in VR will help train doctors and surgeons and test their decision-making and responses to stressful situations in a risk-free way. VR can potentially be used therapeutically too, creating applications to help people cope with anxiety. The technology is also being used to enable consultants based in different locations to collaborate remotely and discuss upcoming surgical procedures. AR glasses can overlay scans and x-rays onto a patient's test results and data, at the bedside, there and then, rather than logging into a desktop computer or checking paper notes (PwC, 2019).

Healthcare is already embracing VR and AR, but so much more is possible. In the coming years, virtual reality will be used more and more to improve the accuracy and effectiveness of current procedures, and enhance the capabilities of the human being, both as the care-giver and the patient. Quite simply, the potential for virtual reality in the healthcare and medicine sector is huge, limited only by the creativity and ingenuity of those creating and applying the technology (Visualise Creative Limited, 2021).

The future of VR lies in its ongoing integration into curricula and with technological developments that allow shared simulated clinical experiences. This will facilitate quality inter-professional education at scale, independent of geography, and transform how we deliver education to the clinicians of the future (Pottle, 2019).

We are now seeing a new computational medical extended reality (XR) concept, which unifies the computer science applications of intelligent reality, medical virtual reality, medical augmented reality and spatial computing for medical training, planning and navigation content creation. It builds upon clinical XR by bringing on novel low-code/no-code XR authoring platforms, suitable for medical professionals as well as XR content creators.

Dr George Papagiannakis of the Department of Computer Science at the University of Crete in Greece, affirms that "virtual/augmented/mixed reality technologies grouped by the industry as XR are now transforming medical training. COVID-19 has forever accelerated advances in this field and elevated these spatial computing technologies from a mere "nice to have" to a "must have" for the medical schools worldwide, the medical device companies and the surgical training centres, that were all challenged by COVID-19 lockdowns. The use of medical VR training is now a reality and part of the "quiet revolution where enterprise VR vastly outnumbers consumer VR applications. Furthermore, healthcare professionals have now acknowledged that training simulators are the number one immediate application of VR/AR in healthcare Intelligence."

Computer technology can create more accurate models and rapidly examine segmentation, feature extraction, classification and tracking of individual cells combining cellular imaging and morphology data such as in non-invasive mapping or modelling of the heart function, evaluation and advancement of therapeutic strategies (Trayanova, 2012; Lefor, 2011).

Computational models excel in modelling complex systems and, consequently, enabling individualisation of medical decision-making and medicine. Computational models are theory- or knowledge-based models, data-driven models, or models that combine both approaches. Data are essential, although to a different degree, for computational models to successfully represent complex systems. The individualised decision making, made possible by the computational modelling of complex systems, has the potential to revolutionise the entire spectrum of medicine from individual patient care to policymaking. This approach allows applying tests and treatments to individuals who receive a net benefit from them, for whom benefits outweigh the risk, rather than treating all individuals in a population because, on average, the population benefits. Thus, the computational modelling-enabled individualisation of medical decision making has the potential to both improve health outcomes and decrease the costs of healthcare (Bukowski et al, 2020).

Computational modelling not only provides a deeper insight of a disease process; novel drug targets may be identified using computational approaches. An essential drug target is identified in a metabolic network of a cell using combined experimental and computational approaches. A successful gene knockout experiment can be modelled *in silico* by deleting a single reaction in the model. Since drugs have multifarious effect on a phenotype, development of a network model is a better approach to understand the effect of a drug on corresponding drug targets and associated side effects (Tiwary, 2019).

However, Wiwanitkit and Wiwanitkit (2015) argue that there are some problems on accuracy. The future presents an opportunity to pay attention to other aspects such as patient-specific modelling (Cueto and Chinesta, 2014).

Although there has been significant progress in the development of computational models for understanding complex diseases during the past decade, there are still many challenges, which need to be addressed by joint efforts of computational biologists and clinicians. The major challenge is to develop a holistic multiscale integrative model capturing emergent property of a complex disease at each scale. Furthermore, a clear understanding of the interaction between different levels of a system will further pave the way for development of novel diagnostics and therapeutics. A patient-specific computational model using data from individual patients will have an added advantage in personalised medicine.

There are benefits and challenges of real-time simulation in the field of computational surgery. The importance of discrete cell models shows that it is possible to utilise descriptive models modified by patient-specific data to further the understanding of the clinical behaviour of diseases or pathologies. The future presents an opportunity to pay attention to aspects such as patient-specific modelling. There are three main barriers to progress in computational medicine based on statistical learning: technological, mathematical, and translational.

There is a principled learning approach for extracting knowledge from large arrays of numbers, and this entails revealing and exploiting disease-related information implicitly stored in high-dimensional, high-throughput biological data. This implies that the traditional approach to biomedical research – which is experimental and molecule by molecule – is not feasible for high-throughput assessment of biological complexity (Winslow et al, 2012).

One of the difficult problems common to virtually all model reduction methods is that of the non-linearity of the problem. Since most living tissues are inherently non-linear, these problems must be solved within the Proper Generalised Decomposition (PGD) framework. In other problems, including physiology in the simulation is simply unavoidable as understanding and quantitatively simulating physiology of the human has been a challenge for some years now. For example, predicting the outcome of vein graft surgery is one of these problems that combine the need for simulation at a macroscopic level, but also at a gene regulatory network level. Simulation at a gene regulatory level is a challenge in itself. The future presents an opportunity to pay attention to other aspects such as patient-specific modelling" (Cueto and Chinesta, 2014).

The models may be improved further so that they can respond to fluctuations in different parameters and environment. The development of a dynamic network model of cancer encompassing initiation, different stages of growth and under metastasis is the need of the hour. A computational model reflecting all connections between all of the neurons (that is, connectome) in the healthy brain is also a priority. It will further help in understanding the dynamic rewiring of the brain under different mental disorders such as major depression, bipolar disorder and schizophrenia (Tiwary, 2019).

The complexity of computational medicine technology, lack of awareness and difficulty of communication among healthcare practitioners may result in lack of or poor acceptance of computational modelling in healthcare and medicine.

Systems for computational medicine can be costly, especially for resource-poor settings such as in low- and middle-income countries.

There is an opportunity, though, because globalisation is making way for joint ventures and foreign direct investments. Although the global economic crisis is a threat to market growth, the computational medicine market is likely to grow due to aging population especially in high-income countries.

It is a challenge that there is lack of regulations or a policy agenda on computational medicine. Where regulations or policies are being planned or being implemented, it is proposed by experts that organisations or institutions that do not comply with standards will be fined. As such, there is need to focus on education and quality standards for both manufactures, providers and consumers in the healthcare and medicine ecosystem. After careful analysis of planned regulations or regulations that are being implemented in certain countries and regions, it is worrying that there is ambiguous liability as there is no case law or no indication of who will be liable if things go wrong with computational medicine. Even if there were to be globally applicable regulations for computational medicine, such regulations would fall short because it is not possible to rationalise as different countries or regions require different formulas.

References

Adapa, K. et al. (2020) Augmented reality in patient education and health literacy: scoping review protocol. *BMJ Open*, 10, e038416.

Albright, G. & McMillan, J.T. (2018) Virtual humans: transforming mHealth for veterans with post-traumatic stress disorder (PTSD). *mHealth*, 4:7.

Al-Mujaini, A. et al. (2011) Satisfaction and perceived quality of an electronic medical record system in a tertiary hospital in Oman. *Oman Med J*, 26(5), 324-328.

Allen, C. (2021) *Extended reality*. Limina Immersive, UK.

Andrews, C. et al. (2019) Extended reality in medical practice. *Curr Treat Options Cardio Med*, 21(18), https://doi.org/10.1007/s11936-019-0722-7.

Banlasadi, T., Ayyoubzadeh, S.M. & Mohammadzadeh, N. (2020) Challenges and practical considerations in applying virtual reality in medical education and treatment. *Oman Medical Journal*, 35(3), e125.

Bohil, C.J., Alicea, B. & Bioccam F.A. (2011) Virtual reality in neuroscience research and therapy. *Nat Rev Neurosci*, 12(12):752-762.

Botella, A. C. et al. (2009) Cybertherapy: Advantages, limitations, and ethical issues. *PsychNology J*, 7(1), 77-100.

Bremmer, R., Gibbs, A. & Mitchell, A.R.J. (2019) The era of immersive health technology. *EMJ Innov*, 4(1), 40-47.

Bruehl, S. & Warner, D.S. (2016) An update on the pathophysiology of complex regional pain syndrome. *PAIN*, 113(3), 713-725.

Bukowski, R. et al. (2020) Computational medicine, present and the future: obstetrics and gynaecology perspective. *American Journal of Obstetrics and Gynecology*, 224(1), pp. 16-34.

Caruso, T.J. et al. () Retrospective review of the safety and efficacy of virtual reality in a pediatric hospital. *Pediatric Quality and Safety*, 5(2), e293.

Claudio, P. & Maddalena, P. (2014) Overview: virtual reality in medicine. *Journal of Virtual Worlds Research*, 7(1).

Chryssanthou, A., Varlamis, I. & Latsiou, C. (2011) A risk management model for securing virtual healthcare communities. *International Journal of Electronic Healthcare*, 6 (2/3/4), 95-116.

Cipresso, P. et al. (2018) The past, present, and future of virtual reality and augmented reality research: a network and cluster analysis of the literature. *Frontiers in Psychology*, 9(2086).

Creswell, K. & Sheik, A. (2013) Organisational issues in the implementation and adoption of health information technology innovations: an interpretative review. *Int J Med Inform*, 82(5), e73-e86.

Cueto, E. & Chinesta, F. (2014) Real time simulation for computational surgery: a review. *Advanced Modeling and Simulation in Engineering Sciences*, 1:11.

Faber, A.W., Petterson, D.R. & Bremer, M. (2013) Repeated use of immersive virtual reality therapy to control pain during wound dressing chamges in pediatric and adult burn patients. *J Burn Care Res*, 34, 563-568.

Field, A. & Butler, R. (2018) *Virtual health: rapid review of evidence and implications*. Waikato District Board. Dovetail.

First Look (2014) *Virtual painkiller*. ASEE-PRISM, Summer.

Fortune Business Insights (2020) Virtual reality (VR) market to touch $120.5 billion by 2026; rapid advancements in deep technology domain to brighten market outlook: Fortune Business Insights.

Garrett, B. et al. (2018) Virtual reality clinical research: promises and challenges. *JMIR Serious Games*, 6(4), e10839.

Gold, J.I. & Mahrer, N.E. (2018) Is virtual reality for prime time in the medical space? A randomised control trial of pediatric virtual reality for acute procedural pain management. *J Pediatr Psychol*, 43, 244-275.

Gold, J.I et al. (2006) Effectiveness of virtual reality for pediatric pain distraction during i.v. placement. *Cyberpsychol Behav*, 9, 207-212.

Grossmann, R.J. (2015) Virtual reality, augmented reality... I call it –Reality. *mHealth*, 1:13.

Hege, I. et al. (2016) A qualitative analysis of virtual patient descriptions in healthcare education based on systematic literature review. *BMC Medical Education*, 16:146.

Hoffman, H.G. (2921) Interacting with virtual objects via embodied avatar hands reduces pain intensity and diverts attention. *Scie. Rep*, 11, 10672, doi:10.1038/s41598-021-89526-4.

Hoffman, H.G. et al. (2019) Immersive virtual reality as an adjunctive non-opioid analgesic for pre-dominantly Latin American children with Large Severe Burn Wounds during burn wound cleaning in the intensive care unit: a pilot study. *Front. Hum. Neuroscie*, 13, 262.

Janani, B. & Somasundaram, M. (2011) *Virtual technology for healthcare: a survey*. Conference Paper, December.

Javaid, M. & Haleem, A. (2020) Virtual reality applications toward medical field. *Clinical Epidemiology and Global Health*, 8, 600-606.

Kassutto, S., Baston, C. & Clancy, C. (2021) Virtual, augmented, and alternate reality in medical education: socially distanced but fully immersed. *ATS Scholar*, 2(4), 651-664.

Kononowics, A.A. (2015) Virtual patients – what are we talking about? A framework to classify the meanings of the term in healthcare education, *BMC Medical Education*, 15:11.

Le May, S. et al. (2021) Immersive virtual reality vs. non-immersive distraction for pain randomized clinical trial protocol. *J. Adv. Nurs*, 77(1), 439-447.

Li, L. et al. (2017) Application of virtual reality technology in clinical medicine. *Am J Transl Res*, 9(9), 3867-3880.

Lluch M. (2011) Healthcare professionals' organisational barriers to health information technologies-a literature review. *Int J Med Inform*, 80(12), 849-862.

Lockwood, D. (2004) *Evaluation of virtual reality in Africa: an educational perspective*. Paris: UNESCO.

Lucas, G.M., et al. (2014) It's only a computer: virtual humans increase willingness to disclose. *Comput Human Behav*, 37, 94-100.

Mandal, S. (2013) Brief introduction of virtual reality & its challenges. *International Journal of Scientific & Engineering Research*, 4(4), 303-309).

Maniatopoulos, S. et al. (2009) Developing virtual healthcare systems in complex multi-agency service settings: the OLDES project. *Electronic Journal of e-Government*, 7 (2), 163-170.

Matamala-Gomez, M. et al. (2018) Decreasing pain ratings in chronic arm pain through changing a virtual body: different strategies for different pain types. *J. Pain*, 20(6), 685-697.

Mohammadzadeh, N. & Safdari, R. (2014) Patient monitoring in mobile health: opportunities and challenges. *Med Arch*, 68(1), 57-60.

Morris, I.D., Louw, Q.A. & Grimmer-Somers, K. (2009) The effectiveness of virtual reality on reducing pain and anxiety in bun injury patients: a systematic review. *Clin J Pain*, 25, 815-826.

Motomatsu, H. (2014) Virtual reality in the medical field. UC Merced Undergraduate Research Journal, 7(1), 207-217.

Mouwad, N. et al. (2020) Augmented realities, artificial intelligence, and machine learning: clinical implications and how technology is shaping the future of medicine. Journal of Clinical Medicine, 9, 3811.

Nichols, S. & Patel, H. (2002) Health and safety implications of virtual reality: a review of empirical evidence. *Appl Ergon*, 33(3), 251-271.

Nunes, F.L. & Costa, R.M. (2008) *The virtual reality challenges in the health care area: a panoramic view.* In: Proceedings of the 2008 ACM symposium on Applied computing; New York: ACM; p. 1312-1316.

Peterson, B.N. et al. (2021) Immersive virtual reality: a safe, scalable, non-opioid analgesic for military and veteran patients. *Frontiers in Virtual Reality*, 2-7, 42290. doi: 10.3389/frvir.2021.742290

Pottle, J. (2019) Virtual reality and the transformation of medical education. *Future Healthcare Journal*, 6(3), 181-185.

PwC (2019) Seeing is believing: how virtual reality and augmented reality are transforming business and the economy.

PwC (2019b) *Changing minds in a changing world*. The future of education in the fourth industrial revolution.

Ratcliffe, J. et al. (2021) Extended reality (XR) remote research: a survey or drawbacks and opportunities. *Proceedings of the 2021 CHI Conference*, 1-13.

Rizzo, A.A., Strickland, D. & Bouchard, S. (2004) The challenge of using virtual reality in telerehabilitation. *Telemed J E Health*, 10(2), 184-195.

Rizzo, A.A., Schultheis, M.T. & Rothbaum, B. (2002) *Ethical issues for the use of virtual reality in the psychological sciences*. In: Bush S, Drexler M, editors. Ethical issues in clinical neuropsychology: Lisse, NL: Swets & Zeitlinger; p. 243-280.

Sampaio, M. et al. (2021) Spanish-speaking therapists increasingly switch to telepsychology during COVID-19; Networked virtual reality may be next. *Telemed. E-Health*, 27, 919-928.

Srivastava, K., Das, R.C. & Chaudhury, S. (2014) Virtual reality applications in mental health: challenges and perspectives. *Ind Psychiatry J*, 23(2), 83-85.

Tiwary, B.K. (2019) Computational medicine: quantitative modelling of complex diseases. *Briefings in Bioinformatics*, 21(2), pp. 429-440.

Tran, J.E. et al. (2021) Immersive virtual reality to improve outcomes in veterans with stroke: protocol for a single-arm pilot study. *JMIR Res. Protoc*, 10(5), e26133.

Trayanova, N.A. (2012) Computational cardiology: the heart of the matter. *International Scholarly Research Network ISRN Cardiology*, Article ID 269680.

van Bennekom, M.J. & de Koning, P.P. (2018) Reducing the stigma on posttraumatic stress disorder in militaries through virtual reality. *mHealth*, 4:5.

Venkatesh, V., Thong, J.Y. & Xu, X. (2012) Consumer acceptance and use of information technology: extending the unified theory of acceptance and use of technology. *Manage Inf Syst Q*, 36(1), 157-178.

Visualise Creative Limited (2021) Virtual reality in healthcare.

Winslow, R.L. et al. (2012) Computational medicine: translating models to clinical care. *Sci Transl Med*, 4 (158), 158158rv11. doi:10.1126/scitranslmed.3003528.

Wiwanitkit, S. & Wiwanitkit, V. (2015) Computational intelligence in tropical medicine. *Asian Pacific Journal of Tropical Biomedicine*, 6 (4), 350-352.

Woolf, C.J. (2011) Central sensitization: implications for the diagnosis and treatment of pain. *PAIN*, 152(3), S2-S15.

Don't miss out!

Visit the website below and you can sign up to receive emails whenever Mbuso Mabuza publishes a new book. There's no charge and no obligation.

https://books2read.com/r/B-A-JPJL-BPSFC

BOOKS2READ

Connecting independent readers to independent writers.

Did you love *Immersive Technologies In Healthcare: Virtual Reality And Augmented Reality*? Then you should read *Health Data Analytics And Informatics*[1] by Mbuso Mabuza!

[2]

With the advent of big data analytics and informatics, and its associated technologies, a rapid growth in career opportunities, coupled with continual technological innovation, has led to an increasing need for qualified professionals who are able to analyse big data and manage technology and information in the healthcare industry.

This book is designed to help you meet that need by providing you with a detailed overview of topics and to explore areas of interest pertaining to data analytics and informatics in the healthcare and biomedical ecosystem. The main topics covered in this book include: data analytics and healthcare informatics; artificial intelligence and

---

1. https://books2read.com/u/bPNl1x

2. https://books2read.com/u/bPNl1x

machine learning technologies; biomedical sensors and trackers; digital clinical trials; high-definition medicine; precision medicine; connected health; how data analytics and informatics can transform healthcare; reflections and perspectives for the future. The book will appeal to people who are interested in the ways in which health data and technology can be used to enhance the quality of health care.

Healthcare informatics is a multi-disciplinary field suited for innovators, healthcare workers, and non-healthcare professionals, united in the goal of improving the quality of health care. Healthcare informatics involves the acquisition, storage and retrieval of healthcare information and aims to ensure the availability of critical data enable the making of sound policies and programmatic decisions to improve patient care across interactions with the health system.

Big Data analytics covers collection, manipulation, and analyses of massive, diverse data sets that contain a variety of data types such as electronic health records (EHRs) to reveal hidden patterns, cryptic correlations, and other intuitions on a Big Data infrastructure Due to its effectiveness, Big Data analytics is widely used in various fields.

Biomedical big data analytics is in the initial adoption phase, and many healthcare organisations want to implement big data analytics to obtain its benefits. To succeed, big data analytics in the healthcare and biomedical ecosystem needs to be packaged so it is menu-driven, user-friendly, real-time and transparent.

Artificial intelligence (AI) and Big Data analytics are seen as novel tools in the planning of health services as well as identifying and monitoring health problems in individuals and populations.

Automation's major benefit is that it helps medical personnel in processing large amounts of patient's data, especially when taking into consideration that medical personnel are often overwhelmed by a series of healthcare tasks. There is potential in delivering more targeted, wide-reaching, and cost-efficient healthcare by exploiting current big data trends and technologies.

The slow pace of innovation in the healthcare industry reflects challenges that are unique to healthcare in implementing and applying sophisticated big data analytics tools, and this points to the need for federal or government policy to emphasise interoperability of health data and prioritise payment reforms that will encourage providers to develop data analytics capabilities.

Salient guidelines and important implications for practitioners and implementers of big data analytics systems that can assist with successful adoption of big data analytics systems in the healthcare system need to be developed. Describing the crucial factors that are required for understanding is important prior to creating a strategy for the acceptance of big data analytics in the healthcare industry, particularly in low- and middle-income countries, where the industry requires filling the gap of big data analytics adoption.

Also by Mbuso Mabuza

A Healthy Mind And Best You: Achieving Great Results in Every
Aspect of Your Life

Purposeful And Better You

Sustainable Development Calls for Effective Strategic Leadership for
Efficient Health Systems

Health Promotion In Low Socioeconomic Settings

Medicine and Sociology of Health

Qualitative Methods In Public Health Research

Global Health Disaster Management

Global Health Policy And Programme Challenges

The Journey of Life Has a Gift of Purpose

Epidemiological Research

Ethics, Qualitative And Quantitative Methods In Public Health
Research

When Love Lasts

Blockchain Technology In Healthcare And Medicine

Immersive Technologies In Healthcare: Virtual Reality And
Augmented Reality

Health Data Analytics And Informatics

How To Improve The Way You Think

Health Systems Engineering: Building A Better Healthcare Delivery
System

Artificial Intelligence In Drug Discovery And Development

How To Make The Most Of Life

Precision Medicine

Connected Health: Technology-Enabled Care

About the Author

Dr Mbuso Mabuza is a highly motivated life-long learner and multi-skilled global health professional. Dr Mabuza's mission is to improve health outcomes and to expand quality healthcare experiences amongst all groups of people and influence change and innovation.